W9-BZV-012

Text-tionary the Ultimate

By Shirley Slee

Text/Symbol De-coder

Illustrations by Jordan Miller,
Rebecca Miller, Gretchen Hess

AuthorHouse™
1663 Liberty Drive
Bloomington, IN 47403
www.authorhouse.com
Phone: 1-800-839-8640

First published by AuthorHouse 7/7/2010

ISBN: 978-1-4520-4725-6 (e)
ISBN: 978-1-4520-4724-9 (sc)
ISBN: 978-1-4520-4723-2 (hc)

Library of Congress Control Number: 2010909806

Printed in the United States of America
Bloomington, Indiana

Acknowledgments

I would like to express my deepest thank you to Donna Sozio, who during our monthly coffee house sessions, helped me turn my dream Text-tionary in to a reality; my husband who is my other half, whether cooking meals, changing diapers, without you, Text-tionary would still be just an idea floating in my head; my parents Jim and Gretchen, who love me without fail, always filling my cup; special thanks to my dear friend Alicia who would tote her cute lil boys over to clean up my writing; my four, the moment I became a mother I knew why I exist! Most of all to my Creator without him nothing is possible, with him all things are!!!

Text-tionary the Guide

The sentence......

Text-tionary

The Ultimate Text/Symbol De-coder

Shirley Michele Slee

(SMS)

Text: A message that is no more than 200 characters. A great way to get the message across without the inconvenience of small talk!

The Case

Text-tionary

This is the ultimate pocket book that will have you sending cool, informative, hip text abbreviations in no time. Text-tionary will also help you de-code over 1,000 different text abbreviations. No more guessing **WDTM** (what does that mean)?

So stretch your thumbs and get ready to learn many new, exciting text abbreviations. Whether you're a first timer, intermediate, or think you're a pro, Text-tionary is for you.

You will learn correct text- etiquette, how not to offend your date, boss, spouse, or friend as well as not sending the ever popular mixed message. One of the most important things to remember when sending a text is MAKE **SURE THAT YOU ARE SENDING IT TO THE CORRECT PERSON :-! (Foot in mouth).**

So, what is texting?

Texting is instant communication. It's fast and efficient, but new abbreviations are being invented every day. Text-tionary decodes any message so you can keep up, and text like you know what you're talking about every time. Even you self-confessed text pros with over 1,000 new abbreviations, smiley faces and numbers, you can't possibly know them all! Try decoding this..... **BY&M, I H8- S/T, GTTP! Thr Txt u SIT**. (Between you and me I hate small talk get to the point! Through text you stay in touch).

The Joys of Texting

There are many joys of texting, first and foremost cutting out the small talk. For example when confirming an appointment it's so easy to text 3:00 or 5:00? Instead of calling and saying "Hello how are you, I was wondering was the appointment 3 o'clock or 5 o'clock? Ok thank you, talk to you later". Another joy of texting is saying goodbye to the "awkward" moment; if you don't know the person that well it is so much easier to communicate through written word, opposed to actual talking, where long pauses can be really uncomfortable. The ability to edit; think about it, when you're fighting with your spouse or have a disagreement with a friend texting allows you to edit, erase, or send. When in person or talking on the phone you don't get the opportunity to take back what was said. One of the best joys of texting is making an announcement like a birth or engagement, ect...

Texting can be very useful, it offers companionship and the promise of connectivity. Over 75 billion texts are being sent every month. Times are changing, **GWTP** (get with the program)...

You see in the subways, crowded restaurants and more when you're surrounded by other people and don't want to disrupt by yapping on your cell phone, but still need to communicate, text is the way to go! It's more discreet.

Uses for Texting

- ❖Keep in touch
- ❖Check in
- ❖Send a quick message
- ❖Confirm a place and/or time
- ❖Cancel a meeting
- ❖Send a funny joke
- ❖Brighten someone's day
- ❖Send little love notes
- ❖Make an announcement
- ❖Send a quick reminder
- ❖Chat for cheap internationally
- ❖Send holiday wishes

And yes, some peeps do use texting to break-up but I don't recommend it.

No pen and paper required. All the information is sent right to your phone. It's an instant written record.

Texting Generation

Not only are teenagers sending more texts, but the 20 something's are sending texts like crazy. Not wanting to be left out the 30 something's are learning how to text as well! And the 40 something's are forced to learn since this is the choice of communication of their tweens. 50 something's are texting because the 20's are!

And before you know it **Gram's** is shooting off texts like a

pro! The number of text messages sent grew nearly three times to a level of ten messages per day per mobile subscriber in the last year.

Fastest texter in the world!

Texting hall of famer

…… Fifteen year old **Kate Moore** from Des Moines, Iowa won the coveted title of fastest texter, as well as a $50,000 dollar prize in the *LG National Texting Championship*. Kate has had a lot of practice sending out about 500 texts a day. During the competition the contestants were blind-folded while running through an obstacle course. Abbreviations were allowed also accuracy and speed was judged. Kate's final text was, Zippity Dooo Dahh Zippity Ayy…My oh My, what a wonderful day!

The Evidence

Texting Vs. Face-to-Face Communication

Everything has its pros and cons. The same goes for Texting and face-to-face communication. When you text you minimize social anxiety and it's a lot easier to ask someone out on a date or for a favor than on the phone. It's also easier to say no to a date or a favor on a text than on a phone. And you don't need to worry about popping a breath mint. When you send a good ol **SWAK**(Sealed with a kiss) through a text.

Through Text you eliminate...

- ❖Social anxiety
- ❖Weird facial expressions
- ❖Reading body language
- ❖Having to shower

- Potential viruses
- STD's
- Leaving the couch
- Awkward silence

Texting, better than voicemail…!

Text messages are instant! The recipient of a text sees a flash on their screen immediately opposed to a voice mail they will not listen to until it's convenient. A voice mail requires one to dial the # then listen, erase, or save, if you must remember a phone number or a list this requires a pencil and paper. With a text you read it right away, and have it for future reference! Also you're more likely to respond within minutes… If you really need to talk, shoot a text with **CM**, meaning call me. This way when it's convenient they will give you a call, or if they are not responding, you can text **LVM** (leaving voice mail). One of the best reasons, you don't have to listen to the stupid recorded voice saying... "If you're not happy with your lame message and want to erase it press... #7. It wastes 25 seconds of your time... x's that by 12 a day, and heck you could have slept in 5 minutes longer.

11 Reasons why!

1. Saves time
2. Faster response time
3. Save important info, for future reference
4. Real time communication
5. Yes or no answers
6. Don't have to stress about how corny your message sounds!
7. Not as confrontational
8. Voice mail is so yesterday!
9. Voice mail was the way our elders sent messages!

10. If you're not keeping up you're falling behind!!

11. Stay in the know while @ office

Text preference

"I would prefer a friend send a text versus leaving a voicemail message any day. I hate voicemail messages but a text you can immediately respond to. Most of the time, all people are saying in the voicemail message is 'Hi, hope you're well. Give me a call'. Now why can't that be said in a text? Quick decisions like where we are going to meet for lunch can also be made via text. Living in the NYC area, I just don't have the time

to stop to listen to messages and time to call back every person who left a message. It's simply unrealistic when I'm catching trains and buses

to get to appointments. I don't sit at a desk all day so texting is the best thing since sliced bread".

Thanks.

Jen

The Benefits of Texting:

- ❖It's fun.
- ❖It's quick and instant communication
- ❖You get your answer right away.
- ❖Allows you to stay in constant contact
- ❖Don't have to spell everything correctly
- ❖Track, I mean check on kids more easily
- ❖You have a hard copy of a conversation
- ❖Great black mail

Look how far we've come!

Smoke messages-

From the past to the future! Texting in the sky, Native Americans were ahead of the game by sending messages through smoke signals. By dosing the fire with grass, or buffalo manure, messages could be sent through the smoke. In general smoke signals were used to transmit news, signal danger, or gather people to a common area.

The Texting Industry...

One-in-four children under the age of eight have a cell phone. This number skyrockets to 89% by the time they reach the age of 11 or 12. 70% of American users sent at least 1 text message every single day. As of June 2008, over 1 trillion SMS messages were sent compared to just 363billion in 2007 and 18 Billion in December 2006. According to the *Nielsen Company* almost 80 messages a day, more than double the average of a year earlier. US Mobile 'Text Message' User Median age **is 38 yrs old.** (And you thought it was just tweens) 49% M / 51% F.

Industry Watch

As of quarter two 2008, a typical U.S. mobile subscriber sends or receives 357 text messages per month, compared to placing or receiving 204 phone calls. Though the number of calls has remained relatively steady, the number of **text messages is up 450%** from just two years prior.

Hot Texting News

Santa Baby, did you get my text???

Every year around Christmas time the Postal Service is bombarded with hundreds of thousands of letters to Santa! In order to save many trees, and not clog the mail service, many cell phone companies have offered a texting option, in which kids can text, their requests to #1224! A random text is sent in return, to ensure the kiddies that Jolly Old Saint Nick got the message.

Cowboys Text too!

"My boyfriend (seems silly to call him that---we're 57---but love makes you feel young!!) was picking me up at work on Friday for a fun night out. I sent him a text stating I was ready. He had been texting me and also his CPA - also a woman - a few times that afternoon. He texted back "OK, baby--I'm comin' to getcha" (we are **Texans**, and, as such don't acknowledge proper spelling or grammar all the time). Of course, this text went to his CPA---a nice lady, but a little reserved. She was a bit taken aback to be approached by her client in that manner. But, in the spirit of understanding, she texted back "You obviously have the WRONG woman!". I'm just glad the message was G-rated....and we are very careful about what we text to each other these days!"

Kelly P. age 57

Texting Highs

When you hear your phone beep that a text came through do you drop everything you're doing and run to the phone to see who it is? If so, you're experiencing a **texting high**. It's like getting a letter in the mail only better because you didn't have to wait two days for it to arrive.

What's your favorite kind of text?

- ❖A picture message
- ❖The kind that keep on firing back ??
- ❖Any kind
- ❖One that doesn't bother you while your chillaxing
- ❖Informative
- ❖Forwards
- ❖Love texts
- ❖Sound texts
- ❖Naughty
- ❖Positive one
- ❖Smiley face with wink
- ❖All abbreviations so I can de-code
- ❖Uplifting quotes
- ❖ I <3 uz
- ❖Joke of the day
- ❖Announcements

Texting Testimonials /Catching the Texting Bug

"I am 98% deaf in one ear, and 42% deaf in the other ear....Text messaging has opened up so much more communication for me. I use to avoid talking on the phone to people, esp. on cell phones, because I struggled so much with hearing all the words that people say! I wing a lot of conversations due to having to constantly ask them to repeat themselves! People get frustrated very quickly with you if you don't hear them the 2nd or 3rd time! I totally

LOVE LOVE LOVE Texting! It has improved my social life so much!"

**Clay* Age 27 Sand Diego, Ca.*

It's so much easier to carry on a good conversation through text messaging. You'll find that you are so much wittier and interesting, through the written word. Rather than having to speak and respond on the spot, you can take a minute to think over your answers! Also, when you don't really feel like talking, you can simply respond to the question instead of having to carry on a ten minute long "polite" conversation!

"Texting has literally changed my life. It's allowed me to keep up with my friends and family who are spread throughout the country. I have a stronger relationship with my cousin and brother because of texting. Before texting, I never really talked to anyone in my family

on the phone besides my mom, grandma, and dad. Living away from everyone, it's hard to keep up. Now that texting is around I know more about their lives than I ever have before. Texting has brought us closer. It also helps me keep up with my old college friends. Texting is a great way to keep up with people between phone calls.

I also use texting for shopping. For example, I signed up thru the Ikea's website and now I get a text message from them any time they're having a sale or a special event. The reason I love texting is because it allows me to live my life and still keep up with my friends and family. That's really important to me".

Jennifer Blanchard

Texts & Emergencies

Text messaging can help families stay in touch in the wake of a disaster, according to a national safety group. *The Safe America Foundation* announced a campaign to train families about alternate ways of staying in touch if traditional communication methods are not working. In the event of an emergency communication may go down, textmessages may still 'get through' or be held in the cue for delivery once service is restored. It is noted that having parents learn how to text message is a valuable safety tool as it may be the only lifeline to their children. The organization's Web site offers a tutorial in emergency text

messaging www.safeamericaprepared.org. **911** texting is particular helpful to the country's deaf and hard-of-hearing residents, who have had to rely on more cumbersome methods to reach **911**.

"As a First Aid instructor, I have been waiting for this technological update for quite some time. I can think of so many instances when someone may be able to text 911 instead of calling! A great example is if you are choking and are unable, to cough, speak, or breathe! If you are not near a land line to call 911, what better way to get some help coming your way? I will be happy to know that at least one person can be saved/helped with this new development. That one life will be worth all the money spent on this new way of reaching a 911 operator."

Chris S.

Saving Africa through Text....

Matt Berg, one of *Times Magazine* 100 most influential people, is looking after 100,000 children, and women in Kenya, Africa! He's the technology director for ChildCount. ChildCount uses texting as a way for health care workers to exchange important health information regarding patient's diagnoses. In only nine months 20,000 nutrition screenings have taken place, 500 cases of malnutrition, as well as 2,000 cases of malaria. Way to go Matt!!!! Just think how much more info the doctors could exchange if they had a Text-tionary.

Verclas Katrin, 2010 Apr 29. Thinkers. Time. <http://www.time.com> Accessed 2010 May 2.

Texting a LOVE STORY

This section could and should and will be a book in it's self! Where do I start, ok when texting it's hard to read what the person is really saying? You can't

read any body language, hear any sighs, or see rolling eyes. For example "I'm coming to town next weekend"! Ok now what type of response do you give? Is this an invite to hang out, or are they just announcing that they will be in town next week, it is really important to communicate what you are trying to say through a text, so finish! Here's an example leaving no guess work, "I'm coming to town next weekend, would u like 2 get together? There, you finished a text, **congrtz**.

Love stories through text...

"Every single night (just about) my boyfriend makes the effort to text me a good night and I love you. It makes me smile every time. And every once in a while, in the middle of the day, he'll text me out of the blue to say, "I love you!" Melissa Zellmer

"My soon to-be wife and I have a great story that almost broke us up early on in our relationship. It had been two months since we started dating and the almighty 3 words (I love you) had yet to cross our lips. It was about 8 in the evening and she had been at a baseball game for a work outing. Rewinding to lunch that day I dropped her off at work, saying our good-byes as she exited the car she said, "pick me up after the game tonight maybe?" I said "sounds great." The game was almost over and I had prepared myself to, cross my fingers, go pick her up. Well, honestly, I was already in the car driving that direction. When I received the text I wanted to see: come pick me up? My twenty minute drive there, was not in vain...until she texted: that was meant for someone else. WHAT?!?!? What the hell is that? Someone else, you are having someone else pick you up from the game!? I abruptly turned the car around and started home. Pissed off, hurt, and confused. 20 minutes later, after locking my car, my phone rang... "Where in the world are you?" Stuck in traffic?" She asked, "I'm at home." I said, annoyed "You haven't left yet" she asked, obviously confused. End of the story, I picked her up.

and we had a great laugh about the miss-communication. Two years later, we're getting married."

Jabez LeBret

Texting Geeks

Are you tired of guessing what **IDK, BRB, TTYL, IDC,** means? It's time to find out! **GWTP** (get with the program) and join in the fun. Text-tionary helps you interpret, understand and join the modern world of Texting. We owe it to the Texting Generation who sends an average of over 2,000 texts per month! The Texting Generation is reinventing the way we communicate. Forget about small talk. Get right down to the point.

Whatever your text is remember you can always add…

AML – all my love!

The Verdict

It's Time to Learn to Text!

Newsflash! If you're not keeping up you are falling behind. Don't be the only one left who doesn't know the difference between **OMG** and **LOL**. We've all received a text leaving us scratching our heads thinking **DKDC** (don't know don't care), but you should care; it won't be the last time you'll receive a text you don't understand and that's the key of good communication. To everyone "holding out" about jumping in the text game, it's not a replacement of heart-to-heart communication only it helps us get through the day faster so we can have more face time. Make an **EF4T** (effort), after reading Text-tionary you will understand why texting is the leader in communication.

"People have been starting to use "y" to mean yes. Unfortunately, it took me a while to catch on. So every time someone would respond to my question with "y", I would be half way through a long explanation about what I was asking before they would simply tell me they already said yes. It's not as simple as the **BRB** or **OMW**, this was way more confusing!"

Marie

"My best friend and I were texting back and forth, talking about my boy troubles! She felt that I should break up with this guy who was treating me badly, not returning my calls, and ignoring me in public. I said I'm going to give him one more chance, she returned the text with **RUS**, I took this as are you stupid! I text back wow you are really understanding! Hurt that she would call me stupid, I stopped texting, she called to say all I said was are you serious! OOPS"

Kellie

How to Use the Text-tionary

It's as **EZ123** to **KW^** (know what's up)! Texting is a genius way to **KIT** (keep in touch). Start first with the abbreviations that pertain to your personality and lifestyle. Most peeps use the same abbreviations over and over again. It's a personal lingo between you and your friends. Soon your friends and family will begin to get used to your abbreviations. They will learn your texting personality and lingo. Save yourself the headache, wondering what does that mean? **GSN** (Get started now)!

Now get acquainted with your phone key pad. Learn how to access your symbols. The symbols are used when making all the different **:-) 's**. If you already know the home row **^5** (high five)! Memorizing the home row helps to return a text at super speed. If you don't have it memorized don't worry, just start getting used to it. It will up your text game.

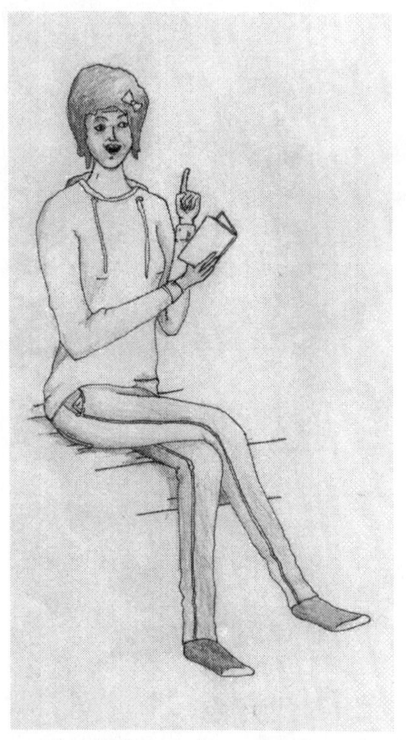

Get to know your texting functions like spell check, auto word maker, symbols... there might be a quick learning curve but it will make texting so much more fun.

Finally what's your texting **:^)** (Personality)? Find your Inner Texter, the way you send a text says a lot about your personality. Are you generally upbeat, sending happy, up lifting quotes? Or are you the bearer of bad news, do you send news of the latest tragedy, sexual predators address, health scares?

Maybe you are a straight shooter, to the point, no funny business, perhaps the class clown, always making people laugh, Mr. Happy Face? Once you find your Inner Texter being shy will be a thing of the past. You can be much more daring in a text. Over time you will develop your own texting personality, dare to be different!

Test Your Texting Personality

Are you:

- **Fast & Furious-** You have the need for speed!
- **Mr. Precise-** You actually spell out the entire message
- **Mr. Monologue-** You show no emotion what's so ever

- **Mr. Exclamation Point!!!!!!!-** You weren't heard enough as a child!
- **Happy Face Overload-** You want people to know how HAPPY you are :-))) (really happy)
- **A Slow poke-** By the time you finish typing the text is irrelevant!
- **Mr. Speedy Gonzales-** You actually see smoke coming from the key pad!
- **The Fwd queen-** You are the one who sends all those corny forwards!

- **The Slacker-** Returning a text is not high on your priority list!
- **Miss Drama Queen-** Every text is due or die!
- **Gloom and Doom-** You send all the latest headlines in the news.
- **Uplifting Quoter-** You send quotes lifting the spirits of friends/ family.

"Sometimes your joy is the source of your smile, but sometimes your smile can be the source of your joy."

Thich Nhat Hanh ☺

Annoying Ambiguous Texts/Double Meanings

Often times an abbreviation can have two or more interpretations such as the famous **LOL**, in the beginning this meant Laugh out Loud, but many started to interpret

LOL as Lot's of Love. There is no right or wrong, you must decipher each text according to its content.

"A friend of mine was with her husband at the hospital. They had gone there because the doctors suspected he had cancer. She had promised several friends that she would let them know what was happening as soon as they found out. She hadn't had experience with texting but thought it would be less disruptive than making a phone call from the hospital. She thought **LOL** meant "Lots of Love" so she sent her friends this message, "It's cancer. **LOL**." After a first round of treatment and what looks like a good long-term prognosis, they can laugh about it now."

Sherri Bergman

Texting 101

One of the benefits of sending a text is that you don't have to spell everything out. Remember, texting is quick messaging, it's not meant to be too long. If it's going to be letter length try sending an email instead.

There's no need to stress if everything isn't spelled out correctly. And there's a foolproof way to make sure you're not a spelling geek. Set your phone to automatically spell-check so you'll know if you made a mistake.

Even though texting is quick, effective instant communication; there's no need to be a stress case about it or be in a rush all the time. Slow down! Chillax! Texting isn't the rat race. Make sure you re-read your texts at least once, but most important that it's sent to the right person. This major mistake has ruined relationships!

Text-iquette Says:

- ❖Respond in due time
- ❖Have something to say
- ❖Use "haha's" generously
- ❖Try to continue the conversation
- ❖If the conversation is dying let it die
- ❖If you have way too much to say just call

Texting Don'ts

Don't Spam Your Friends

Don't sell out your friends! **DEGT** (don't even go there) If you've got nothing but time to send, "tell ten of your closest friends how much you love them in the next 5 minutes, or you'll live a lonely life," texts, please take up a **new hobby**. Send cool text abbreviations that will have them reaching for their Text-tionary to de-code the message! **DWMT** (don't waste my time) with spam!!!

Here's a sample of a lame forward I received the other day. "I am your......." fill in the blank & then pass it on n c how many crazy answers u get. Describe me in 1 word just 1 & send it 2 me; then send this msg. to 20 people & see what you get back. Finish this sentence "I have always wanted to tell u that......... text me the answer and then pass it on."

"A peach is a peach, a plum is a plum, a kiss aint a kiss unless it's done with tongues, so open your mouth and close your eyes and give your tongue some exercise... Keep this going!"

So lame.......

Make a Wish "Believe me.. This really works!! Whatever age you are, is the number of minutes it will take for your wish to come true (ex. you are 15 yrs.old = 15 minutes). If you don't send this to 20 people in 20 minutes, you will have bad luck for 20 years!! GO!!!" **Really????**

Don't Text & Walk

A Staten Island girl fell into an open man hole while texting, suddenly the ground beneath her disappeared! Thankfully she only suffered only minor cuts and scratches, it could have been much worse. Also Lampposts in Britain on Brick Lane have been wrapped in cushions, to protect walk'n'texters from hurting themselves! LOOK UP WHEN WALKING!!

Don't Text & Drive (DWT) *Driving while texting*

Currently, texting while driving is banned in 23 and counting states, including Alaska, California and New Jersey, as well as the District of Columbia. Utah passed the nation's toughest law to crack down on texting behind the wheel. Offenders now face up to 15 years in prison. Studies show that talking on a cell phone while driving is as risky as driving with a .08 blood alcohol level — generally the standard for drunk driving — and that the risk of driving while texting is at least twice that dangerous. Driver distractions (like texting) are a factor in 25–50 percent of all crashes. Car accidents are already the number-one killer of teens (ages 15–20) in the United States, claiming over 6,000 lives every year. What about those sleepy teens who talk all night and text while driving? Practice putting your phone in the glove compartment while behind the wheel, that way you're not tempted to check that text!! Justin Bieber signed Oprah's No Phone Pledge, what about you???? Well if not, you'll have a chance to take Shirley's No Phone Zone Pledge.

List of big names who have signed O's no phone zone pledge so far.....

Brooke Burke

Derek Hough

Diane Sawyer

Dr. Oz

Jada Pinket smith

Jeff Bridges

Jerry Seinfeld

Justin Bieber

Lisa Ling

Mary J. Blige

Usher

Will Smith

Mo'Nique

Morgan Freeman

Queen Rania of Jordan

Randy Jackson

Raquel Welch

Sandra Bullock

Shaun White (Olympics)

Tina Fey

Tom Cruise

Tyler Perry

Take Shirley's No Phone zone pledge

Shirley's No Phone Zone Pledge

I pledge to make my car a No Phone Zone. Beginning right now, I will do my part to help put an end to distracted driving by committing to drive as responsibly as I can:

- I Will not text while I am driving
- I will not text while driving and will use only hands free calling if I need to speak
  on the phone while I am driving.
- I will not text or use my phone while I am driving. If I need to use my phone, I will
  pull over to a secure location.

____

Name   Date

Don't Over Punctuate

Sometimes people tend to over express themselves through text like over using capitals, or the infamous Mr. Exclamation point!!!! Can't forget the overuse of bold letters we get it you really mean it! Please don't forget when you are fighting with someone to get rid of the smiley face at the bottom, "I can't stand u, you are such a jerk!" **:-)))** (Very Happy)

Don't Take it Personally

It can be difficult to decode a text when you don't have the advantage of reading body language, or hearing a sigh on the phone. For example when you have to cancel an appointment through text and the response is **K**. Instantly you wonder oh great did I upset them, or are they fine with it? One liners are another text that is hard to decode for example:

"Coming over?" Am I coming over, are you coming over, who's coming over? A good response to a one liner is, "this a **?**."

Don't Jump to Conclusions

Story of Texting Dad trying to be cool!

"My 15-year old son texted me to ask permission to go to a friend's house after the movie let out. Wanting to be cool, like him, I texted back with a single letter: Y. I intended it to mean Yes, as in Y/N. He took it as "why?" and rather than hassle with explaining it to me, he came directly home."

Harry Liebman

What's the F.P.S. I of this couch?

It's basically a done deal, nothing left but to take it home. I've hijacked my brother in-law's truck, wrangled my friend as co-pilot, body guard. MapQuest directions in hand were on our way. For a couch that was being sold as less than two years old, this thing looked worked! It showcased pancaked cushions that were still warm with fur. A "Hand Wash Only" slip cover that appeared to have been over baked in some dysfunctional dryer then wrestled back on. All the while, I combed over the couch doing the math. I hear several, dogs wolfing

loudly as they were feverishly gnawing at the back door, just aching to resume their standard molded positions on this Pottery Barn "Style" dog bed, I mean "couch"! I discretely motioned to my bud this was a "no-go" and let's get the heck out of here. Then I informed the seller that the deal was off, due to misrepresentation. Upon that information she became nasty. Those wild garaged dogs would be a welcome exchange for this ferocious female that was starting to stew in her own juices. My body guard quickly steps in with a few positive comments, he wasn't getting anywhere. He then asks the all important question. "What would you guess the **F.P.S.I.** is on this thing, lady?" She goes, "What!??" "**F.- P-. S-. I**, as in FARTS Per Square Inch!" That's it!! We quickly booked it out of there before she let those dogs loose! *Lynda*

Don't Get Addicted

Confessions of a Texting Addict (almost like a druggie)

"All my friends know that I'm addicted to my phone, especially for texting. I really can't breathe without it! I rely on texting since I'm in a long-distance relationship – it's our way to catch up without having to constantly play phone tag. It's nice to know he's just a text away and that he's thinking of me, even if we are 300 miles away from each other."

Veronica, FEMALE, 27, Austin, Texas

Don't Drink & Text

TFLN (txts frm lst nght)

This is when you wish you had the power to go into some one else's phone and erase the texts you sent last night!!! You were fighting and now you've made up. You said some pretty nasty things, you wish you could erase, down fall of texts **it is written**! You were impaired, drunk, **DUR** (do you remember) and sent some pretty stupid texts, now the receiver has some great black mail on you! Oh well **W/E**....(whatever)

Don't Get Busted!

Deleted but not dead! With all this written word, floating around Texters beware. Like e-mail and Internet instant messages, text messages tend to be saved on servers. “One of the false assumptions that people make is that when they hit the delete button, messages are gone forever, but nothing can be further from the truth," said Jeff Kagan, an independent telecommunications analyst in Atlanta. In Europe and Asia, where texting is hugely popular, some criminal cases have hinged on them. "Don't ever say anything on e-mail or text messaging that you don't want to come back and bite you."

Peter Pan Exposed

"Late last year, banking intern Kevin Colvin texted his boss saying that he had a family emergency that required him to return home. His initially understanding boss wasn't very happy when his attention was drawn to a photo on Kevin's Face book account at the time he was supposed to be

tending to a sick relative. The fact that he was dressed liked, Peter Pan – at a costume party, Added insult to injury, and he was promptly let go from his job." *Tom Allys*

Texting dangerously.... *(Can anyone say Tiger?)*

Blackmail through texting is increasing in popularity, now it not just “he, said”, “she, said”, its “he, texted”, “she, texted”! Unfaithful spouses in France, passionate text messages sent to mistress and lovers can now be used as evidence against you in a divorce. Previously, husbands and wives often had to wait for years to escape a marriage if they could not prove that their spouse was misbehaving or mistreating them. So be confident that the texts you are sending can’t be used against you in a court of law!

Ooops!!!!

“Going through my divorce I started dating another guy, when I was texting the new guy one night, I accidently sent a naughty text to my ex!

He responded, I know that wasn't meant for me!! I felt terrible....."

Aprylle Taylor

How to Extinguish the uber-frequent texter:

***Text trap*!** You've opened a floodgate! In the event of a "text trap," which is when someone keeps texting you and won't stop, do not try to get out of it simply by saying "**haha**," "**lol**," or "yeah." The person may either not get the hint or continue texting you, or they may feel that you are a rude texter and not wish to associate with you. The best thing to do is to simply stop responding. They will understand. This also works when a friend says something you appreciate, but have no response for. Just let the conversation go.

Texting Rudeness

DBAJ (Don't be a jerk)! The infamous texting while at the dinner table, no one appreciates this. Besides just being flat out rude, this is a sacred time to enjoy each other's company,

listen, talk, and engage! You're not fooling anybody! When talking with one person, give them your full attention please. Everybody knows you can't do two things at once. Carrying on a conversation verbally and electronically does not work. Someone feels left out. The message that you're sending to the live person in front of you is that, "yeah I'm half listening, and you're really not that important for me to give you 100% of ME". So be careful!

This reminds me when my recently divorced cousin came to visit, I swear only half of her really was there, she literally texted the whole entire time she "visited". After two days of tolerating her texting, while "talking" to me, eating meals, even on a five mile walk! I had enough and threatened to throw her phone out the window while driving, she got the message.

***Do not text*......**

- ❖During meals
- ❖Your suppose to be listening to someone

- ❖While you're telling a story
- ❖When driving
- ❖Checking out at a register
- ❖Ordering food
- ❖At Starbucks waiting in line
- ❖In the movies (we see the glowing light)!
- ❖In a place of worship
- ❖At child school performance
- ❖During a wedding
- ❖While at the gym
- ❖On a date
- ❖In class
- ❖Walking on a busy side walk
- ❖At the park (Quality time with your child)!
- ❖While at the opera
- ❖On a guided tour
- ❖While at the day spa

If you must text during any of these events, please don't forget to turn off your text alert noise, whether it be a BEEP, music, or some other annoying noise.

Texts Gone Wrong: How to fix it & What to Do!

The first step is **prevention**.

Hot Tip #1: Keep your phone book organized with symbols or descriptions that *mean something to you. Having your contacts organized will reduce the mistake of sending the wrong person a message. For example if you've dated people with the same name put something that reminds you of that person next to their name, for example John (lot's of muscles), John (skinny , funny).....

"I started texting this girl while I was out the other night, (so I thought) I met her and a few others at a club. Finally we agreed to meet up and it wasn't the girl I thought it was! I mixed the girls up in my phone and texted the wrong one."

**John* 25, Hoboken, New Jersey*

Hot Tip #2

If a texts lands in some one else's inbox on accident, the best thing to do is, once mistake is realized quickly apologize "**OOPS S2WP SRY** (Sorry to wrong person)…. Beware this can be an extremely embarrassing situation.

"I still shudder to remember mine. I was frustrated with something my friend said on text and I wrote a bad word on the reply and then wrote a sentence starting with "you". Unfortunately I was in a

hurry and forgot the period between the expletive and the "you". He thought I said "F you". He called me a name back on text and I tried to explain to no avail. I even called and he wouldn't pick up his phone. He didn't speak to me for two days. I was SO upset". *Julie*

"I am 25, great with technology, might be called an early adopter, been texting for 10 years by now... And last night I sent the wrong person the wrong text. It should have gone to my mom, but instead went to a cute blond girl I've been talking to. We haven't know each other long... I was texting her and my mom at the same time". The text was: "I Love you too, Niiiight!" Oops! *Evan*

“My husband and I on our 2nd marriage and have 6 boys together 4 of which live with us full time. The opportunity to get alone time is far and few between. That being said my husband and I got to go out and get a quick bite. When we got home we had a little time before we had to get my middle son who is 14 from basketball practice. So we were making out. That's all we were doing was kissing. I looked at the clock and said you gotta go get Scott. Reluctantly my husband left. I felt I would give him a little love and decided to text him. I texted, "I sometimes forget how attracted I am to you and how much I love kissing you." About a minute goes by and I receive a response, the text says, "mom ????" OMG I started hysterical laughing and I texted my son back "awkward" at the same time my husband texted me "awkward" I sent it to my son by mistake"..

Elise

"I'm working out with a trainer now. I was so happy one morning when I lost some weight that I texted him my weight, but I didn't text him, I texted a co-workers wife (one up on my contacts tab). She asked her husband if I wanted her to text "her" weight back (she was dieting too)??? She didn't know what to do... Very embarrassing"!

Jennifer

Texting for Dollars:

How Texting Can Save You Money

Mycoupster.com, get savings via text! Walk into a store, call in and receive your coupon...it's awesome. Have stores send you notifications when they are having sales.....

Your bank can send you notification of account activity. Say you're account hit's a low balance, your bank will send an alert through text.... The best part, no more overdraft charges!!!!!!

Texting and T.V

American's next Idol is in the palm of your hand, literally! By casting a vote through text you can help choose the next *American Idol*......

Dancing with the Stars, is another show where you're text has the power to crown a winner, watching dancers Tango their way to victory, your vote counts!

Texting.... OVER IT! When it's too much...!

When you dread hearing your phone ring and actually get heart palpitations when you press the answer key, your faced with a live voice, it's time to start facing the fact that you've developed a social anxiety about having to actually talk to someone.

What texting is not a substitute for … hugs… face time, a real person… texting gives you more time because it saves time.

Text & Have it All

You can have all the benefits of texting… plus a real life. With the amount of time you save through text, you ultimately have more face time! Texting is a tool with personality… what's yours? **TT4N**

Make your own abbreviations, that only you and your friends know-

Space for new texts...

____

____

____

____

____

____

____

____

____

____

____

____

Text-tionary (Letters)

Slang: texting abbreviations. A necessity when texting since there are only 160 characters to be used, one must get across as much as possible! You're not graded on punctuation, grammar, and capitalization. Words that have no abbreviation, texter's remove the vowels from a word, thus leaving a string of consonants to be interrupted. For example "down" becomes "**DWN**", or "Enough" becomes "**ENUF**".

LETTERS

A

AAK- Alive and kicking

AAP- Always a pleasure

AAR- At any rate

AAS- Alive and smiling

ABT- About

ADBB- All done, bye bye

ADR- Address

AFC- Away from computer

AH- At home

AIGHT- Alright

AITR- Adult in the room

AMAP- As much as possible

AML- All my love

AMOF- As a matter of fact

AND- Any day now

AOTA- All of the above

APAC- All praise and credit

ASAP- As soon as possible

ATB- All the best

ATEOTD- At the end of the day

ATM- At the moment

AWC- After while crocodile

AWESO- Awesome

AYDY- Are you done yet

AYOR- At your own risk

AYS- Are you serious

AYT- Are you there

B

B- Back

B/C- Be cool

B/F- Boyfriend

B/G- Background

B4- Before

B4N- Bye for now

BAK- Back at keyboard

BAU- Business as usual

BAY- Back at ya

BB- Be back

BBIAF- Be back in a few

BBL- Be back later

BBN- Bye bye now

BBQ- Barbeque

BBS- Be back soon

BC- Because

BCY- Be seeing you

BD- Big deal

B-Day- Birthday

BF- Best friend

BFF- Best friend forever

BFFL- Best friends for life

BFN- Bye for now

BHL8- Be home late

BI5- Back in five minutes

BIF- Before I forget

BIL- Boss is listening

BIL- Brother in law

BION- Believe it or not

BL- Belly laugh

BLNT- Better luck next time

BO- Back off

BOL- Best of luck

BRB- Be right back

BRD- Bored

BRT- Be right there

BTA- But then again

BTDT- Been there done that

BTT- Back to topic

BTW- By the way

BWL- Bursting with laughter

BY&M- Between you and me

BYTM- Better you than me

C

C&G- Chuckle & Grin

CB- Coffee break

CFI- Crazy for it

CID- Consider it done

CLAB- Crying like a baby

CM- Call me

CMB- Call me back

CMIIW- Correct me if I'm wrong

CMON- Come on

Congrtz- Congratulations

CR8- Create

CRB- Come right back

CRBT- Crying really big tears

CRS- Can't remember stuff

CSL- Can't stop Laughing

CT- Can't talk

CTC- Care to chat

CU- See you

CUA- See you around

CUL- See you later

CWOT- Complete waste of time

CWYL- Chat with you later

CYE- Check your e-mail

CYO- See you online

D

D/L- Down load

DBAJ- Don't be a jerk

DBAU- Doing business as usual

DC- Disconnect

DD- Dear daughter

DEGT- Don't even go there

DH- Dear husband

DIKU- Do I know you?

DIY- Do it yourself

DKDC- Don't know don't care

DL- Down low

DLTBBB- Don't let the bed bugs bite

DM- Doesn't matter

DNR- Dinner

DNT- Don't

DNU- Do not understand

DQMOT- Don't quote me on this

DR- didn't read

DS- Dear son

DTRT- Do the right thing

DTS- Don't think so

DUR- Do you remember

DW- Dear wife

DWMT- Don't waist my time

DWN- Down

DWT- Driving while texting

DXNRY- Dictionary

DYNWUTA- Do you know what you are talking about

E

E1- Everyone

E2EG- Ear to ear grin

EAK- Eating at keyboard

EF4T- Effort

EM4BI- Excuse me for butting in

EMA- Email address

EMSG- Email message

ENUF- Enough

EOD- End of day

EOD- End of discussion

EOL- End of lecture

EOM- End of message

ES- Erase screen

ETA- Estimated time of arrival

EV- Ever

EZ123- Easy as one, two, three

EZY- Easy

F

F2F- Face to face

FBF- Fat boy food

FBM- Fine by me

FC- Fingers crossed

FCOL- For crying out loud

FICCL- Frankly I couldn't care less

FIMH- Forever in my heart

FIIOOH- Forget it I'm out of here

FITB- Fill in the blank

FOTB- Father of the bride

FOTG- Father of the groom

FPSI- Farts per square inch

FRT- For real though

FTBOMH- From the bottom of my heart

FW- Forward

FWIW- For what it's worth

FWM- Fine with me

FYA- For your amusement

FYEO- For your eyes only

FYI- For your information

G- Giggle

G/F- Girlfriend

G2CU- Good to see you

G2G- Got to go

G2R- Got to run

GA- Go ahead

GAL- Get a life

GAS?- Got a second?

GB- Good bye

GBTW- Get back to work

GBU- God bless you

GFI- Go for it

GFN- Gone for now

GFTD- Gone for the day

GG- Gotta go

GGMSOT- Gotta get me some of that

GGOH- Gotta get out of here

GIAR- Give it a rest

GJ- Good job

GL- Good luck

GL/HF- Good luck have fun

GMTA- Great minds think alike

GN- Grin

GNIGHT- Good night

GNSD- Good night sweet dreams

GOI- Get over it

GOL- Giggling out loud

GR8- Great

GRATZ- Congratulations

GRL- Girl

GRWG- Get right with God

GS- Get serious

GSN- Get started now

GT- Good try

GTCU- Great to see you

GTG- Got to Go

GTTP- Get to the point

GUD- Good

GWTP- Get with the program

H

H- Hug

H&K- Hug and kiss

H/W- Home work

H2CUS- Hope to see you soon

H8- Hate

HAG1- Have a good one

HAGN- Have a good night

HAGO- Have a good one

HAND- Have a nice day

HAU- How about you?

HB- Hurry back

H-BDAY- Happy birthday

HBU- How about you?

HF- Have fun

H-FDAY- Happy fathers day

HHIS- Head hanging in shame

HLO- Hello

H-MDAY- Happy mothers day

HOAS- Hold on a second

HRU- How are you?

HTH- Hope this helps

HV- Have

I

I2- I too

IA8- I already ate

IAC- In any case

IAE- In any event

IAO- I am out

IAT- I am tired

IB- I'm back

IC- I see

ICAM- I couldn't agree more

ICDI- I can't discuss it

IDC- I don't care

IDK- I don't know

IDTS- I don't think so

IDUNK- I don't know

IG2R- I got to run

IHNI- I have no idea

IK- I know

IKR- I know right!

ILB8- I'll be late

ILU- I Love you

IMHO- In my humble opinion

IMO- In my opinion

IMS- I am sorry

IMSB- I'm so bored

IMU- I miss you

IOMH- In over my head

IOW- In other words

IRL- In real life

ISLY- I still love you

ITM- I'm the man

ITUN- I though you knew

IUKWIM- If you know what I mean

IUSS- If you say so

IWALU- I will always love you

IWO- I want out

IYO- In your opinion

IYSS- If you say so

J

J4F- Just for fun

JAM- Just a minute

JAS- Just a second

JGI- Just Google it

JIC- Just in case

JJA- Just joking around

JK- Just kidding

JLMK- Just let me know

JMO- Just my opinion

JP- Just playing

JTLUK- Just to let you know

JW- Just wondering

K

K- Okay

KFU- Kiss for you

KIA- Know it all

KISS- Keep it simple silly

KIT- Keep in touch

KK- Knock, knock

KK- Okay, Okay

KOC- Kiss on cheek

KOL- Kiss on lips

KUGW- Keep up the good work

KW^- Know what's up

KWIM?- Know what I mean?

L

L2C- Like to come

L2G- Like to go

L8R- Later

LBAY- Laughing back at you

LD- Later dude

LHM- Lord help me

LIC- Like I care

LMBO- Laughing my butt off

LMIRL- Let's meet in real life

LMK- Let me know

LOL- Laughing out loud

LOL- Lots of love

LOTI- Laughing on the inside

LQTM- Laughing quietly to myself

LTD- Living the dream

LTNS- Long time no see

LTS- Laughing to self

LVM- Leaving voice mail

LYA- Love Ya

LYLAS- Love you like a sis

LYLC- Love you like crazy

LYSM- Love you so much

M

MC- Merry Christmas

MFI- Mad for it

MGBY- May God bless You

MIRL- Meet in real life

MKAY- Meaning okay

MNC- Mother nature calls

MOB- Mother of the bride

MorF- Male or Female

MOTG- Mother of the groom

MSG- Message

MSU- Make stuff up

MTE- My thoughts exactly

MTFBWY- May the force be with you

MUSM- Miss you so much

MYOB- Mind your own business

N

N1- Nice one

N2M- Nothing 2 much

NA- No access

NALOPK- Not a lot of people know

NANA- Not Now No need

NBD- No big deal

NBFAB- Not bad for a beginner

NC- Nice crib

NC-No comment

NE- Any

NE1- Any one

NFM- Not for me

NFS- Not for Sale

NFW- Not for work

NIGI- Now I get it

NLT- No later than

NM- Nothing much

NMH- Nothing much here

NMP- Not my problem

NMU?- Nothing much, you?

No1- No one

NOYB- None of your business

NP- No problem

NRN- No response necessary

NSFW- Not safe for work

NSISR- Not sure if spelled right

NT- Nice try

NTHING- Nothing

NTL- Not till later

NVM- Never mind

NVR- Never

NW- No way

NWO- No way out

O&O- Over and out

O4U- Only for you

OB- Oh baby

OB- Oh brother

OIC- Oh I see

OJ- Only joking

OL- Old lady

OM- Oh my

OM- Old man

OMDB- Over my dead body

OMG- Oh my gosh

OMW- On my way

OOH- Out of here

OOPS- Totally messed up

OOTD- One of these days

OOTO- Out of the office

OP- On phone

ORLY- Oh really?

OT- Off topic

OTB- Off to bed

OTFL- On the floor laughing

OTL- Out to lunch

OTOH- On the other hand

OTP- On the phone

OTRC- On the red carpet

OTT- Over the top

OTTOMH- Off the top of my head

OTW- Off to work

OU- Owe you

OVA- Over

OYH- Over your head

P

P2P- Parent to parent

P2P- Peer to Peer

P911- Parents are in the room alert

PAW- Parents are watching

PCM- Please call me

PDA- Public display of affection

PDH- Pretty darn happy

PDQ- Please don't quit

PDQ- Pretty darn quick

PDS- Please don't shoot

PEEPS- People

PIC- Picture

PIP-Peeing in pants

PIR- Parents in room

PISS- Put in some sugar

PITB- Pain in the butt

PL8- Plate

PLMK- Please let me know

PLS- Please

PLU- People like us

PLZ- Please

PM- Private message

PMFI- Pardon me for interrupting

PMFJI- Pardon me for jumping in

POAHF- Put on a happy face

POS- Parent over shoulder

POV- Point of view

POV- Privately owned vehicle

PPL- People

PPU- Pending pick up

PRT- Party

PSOS- Parent standing over shoulder

PT- Pretty tight

PTL- Praise the Lord

PTMM- Please tell me more

PU- That stinks

PUKS- Please pick up kids

PXT- Please excuse that

PZ- Peace

PZA- Pizza

Q

QFE- Questions for everyone

QFI- Quoted for irony

QIK- Quick

QL- Quit Laughing

QQ- Quick question

QT- Cutie

QTPI- Cutie Pie

R

R8- Rate

RBAY- Right back at you

RIP- Rest in peace

RL- Real life

RLY- Really

RME- Rolling my eyes

RMLB- Read my lips baby

RMM- Reading my mail

ROTFL- Rolling on the floor laughing

RTM- Read the manual

RTQ- Read the question

RU- Are you?

RUOK- Are you ok?

RUS- Are you serious

RUT- Are you there?

RYB- Read your Bible

RYS- Are you single?

RYS- Read your screen

S

S/T- Small talk

S1WM- Someone with me

S2R- Send to receive

S2S- Sorry to say

S2WP- Sent to wrong person

SAHD- Stay at home dad

SAHM- Stay at home mom

SAL- Such a laugh

SBT- Sorry about that

SC- Stay cool

SD- Sweet dreams

SDMS- Sweet dreams my sweetie

SETE- Smiling from ear to ear

SFAIK- So far as I know

SH- Same here

SH^- Shut up

SICNR- Sorry I could not resist

SIH2R- Sorry I have to run

SIMUC- Sorry I missed your call

SIT- Stay in touch

SK8- Skate

SK8NG- Skating

SK8R- Skater

SK8RB- Skater boy

SK8RG- Skater Girl

SLAP- Sounds like a plan

SME1- Some one

SMHID- Scratching my head in disbelief

SO- Significant other

SOL- Sooner or later

SOMY?- Sick of me yet?

SorG?- Straight or Gay?

SOS- Help

SOTMG- Short of time must go

SPK- Speak

SPST- Same place same time

SPTO- Spoke to

SQ- Square

SRSLY- Seriously?

SRY- Sorry

SSDD- Same stuff different day

SSRY- So sorry

ST&D- Stop texting and drive

STR8- Straight

STW- Search the web

SUITM- See you in the morning

SUL- See you later

SUL- See you later

SUP- What's up

SUS- See you soon

SUX- Suck's

SWAK- Sealed with a kiss

SWDYT-So what do you think?

T

T+- Think positive

T2Go- Time to go

T2UL- Talk to you later

T4BU- Thanks for being you

TAL- Thanks a lot

TA:M- Tomorrow am

TA4N- That's all for now

TAU- Thinking about you

TBC- To be continued

TBD- To be determined

TBH- To be honest

TBL- Text back later

TC- Take care

TCB- Take care of business

TCOY- Take care of yourself

TD2U- Totally devoted to you

TFS- Thanks for sharing

TGIF- Thanks God it's Friday

THNQ- Thank you

THR- Through

THT- Think happy thoughts

THX- Thanks

TIA- Thanks in advance

TIAD- Tomorrow is another day

TIC- Tongue in cheek

TILII- Tell it like it is

TISC-That is so cool

TL- Too long

TM2H- To much to handle

TMB- Text me back

TMI- To much info

TMIY- Take me I'm yours

TMOT- Trust me on this

TMW4I- Take my word for it

TNT- Till next time

TOJ- Tears of joy

TOU- Thinking of you

TP:M- Tomorrow PM

TPTB- The powers that be

TRBL- Trouble

TSNF- That's so not fair

TSTB- The sooner the better

TT4N- Ta Ta for now

TTLY- Totally

TTTT- These things take time

TTUL- Talk to you later

TTUS= Talk to you soon

TU- Thank you

TU- Thank you

TUSM- Thank you so much

TUVM- Thank you very much

TWD- Texting while driving

TWSS- That's what she said

TYS- Told you so

TYT- Take your time

U

U- You

U4E- Yours for ever

UCMU- You crack me up

UCRTN- Uncertain

UDM- You da man

UFN- Until further notice

UGTBK- You got to be kidding

UL- Upload

UN4U8- Unfortunate

UNPC- Un politically correct

UOK- Are you ok?

UR- You are

URA*- You are a star

URH- You are hot

URSK2M- You are so kind to me

URT1- You are the one

URTM- You are the man

URW- You are welcome

USBC- Until something better comes

UT2L- You take to long

UW- Your Welcome

VBS- Very big smile

VEG-Very evil smile

VF- Very funny

VG- Very good

VGC- Very good condition

VIP- Very important person

VM- Voice mail

VN- Very nice

VRY- Very

VSF- Very sad face

W

W/- With

W/B- Write back

W/E- What ever

W/O- With out

W@= What?

W2G- Way to go

W4U- Waiting for you

W8- Wait

WAH- Working at home

WAJ- What a jerk

WAM- Wait a minute

WAN2- Want to?

WAN2TLK- Want to talk

WAS- Wait a second

WAU?- What about you

WAU@?- Where are you at?

WAUF?- Where are you from

WAWA?- Where are we at

WB- Welcome back

WBS- Write back soon

WC- Welcome

WC- Who cares

WCA- Who cares anyway

WDALUIC- Who died and left you in charge

WDUK?- What do you know?

WDUM- What do you mean?

WDUT?- What do you think?

WDUW- What do you want?

WH5- What, who, where, when, why

WIBNI- Wouldn't it be nice if

WIIFM- What's in it for me?

WITP- What is the point?

WITW- What in the world?

WIU- Wrap it up

WK- Week

WKD- Weekend

WOM- Waste of money

WOMBT- Waste of money, brains, time

WP- Wrong person

WRK- Work

WRT- With regard to

WRUD- What are you doing?

WRUF?- Where are you from?

WRUU2- What are you up to?

WTG- Way to go

WTH- What the heck?

WTM- Who's the man?

WU- What up

WUF?- Where you from?

WUGAM- When you get a minute

WULEI- When you least expect it

WUMM-Will you marry me?

WUSIWUG- What you see is what you get

WUWH- Wish you were here

WWJD-What would Jesus do?

WWUC- Write when you can

WWUCM- When will you call me?

X- Kiss

XLNT- Excellent

XME- Excuse me

XOXO- Hugs and kisses

Y

Y- Yawn

Y?- Why

Y2K- Your to kind

YARLY- Ya, really

YBIC- Your brother in Christ

YBS- You'll be sorry

YCHT- You can have them

YCLIU- You can look it up

YCMU- You crack me up

YF- Wife

YG- Young gentleman

YG2BKM- You've got to be kidding me

YGG- You go girl

YHBH- You have been had

YHBW- You have been warned

YHL- You have lost

YHLI- You have lost it

YIU- Yes I understand

YKW- You know what

YKWUCD- You know what you can do

YL- Young lady

YNK- You never know

YR- Yeah right

YR- Your

YSIC- Your sister in Christ

YSUD- Yeah sure you do

YT- You there?

YTG- You're the greatest

YTTL- You take to long

YW- You're welcome

YWSULS- You win some you lose some

Z

Z- Zero

Z%- Zoo

ZH- Sleeping hour

ZOT- Zero tolerance

ZUP- What's up

Z^- What's up

ZS- Z's are calling (Sleep)

ZZZ- Bored/ Tired

Text-tionary (Symbols)

The smiley can be confusing, but it has lots of meanings, as you browse through the symbol section pay close attention as there are many cute, silly, informative ways to express just about anything! Who knew **I:-O** =Flat Top Loudmouth ?

Symbols/Smiley

@>--;-- - =A rose

O:-) =Angel

0*-) =Angel wink - female

0;-) =Angel wink - male

:-Z =Angry face

:-{{ =Angry Very

>:-(=Annoyed

~:o =Baby

@:-} =Back From Hairdresser

~~8-O =Bad-Hair Day

d:-) =Baseball

:-) =Basic

:-{0 = Basic Mustache

:~-(=Bawling

:-){ =Beard

(:-{~ =Beard - long

: = =Beaver

%-| =Been up All Night

:-)^< =Big Boy

(:-) =Big Face

:-)8< =Big Girl

(((H))) =Big Hug

:-X =Big Wet Kiss

=|:o} =Bill Clinton smiley

(:-D =Blabber Mouth

?-(=Black Eye

(:- =Blank Expression

#-) = Blinking

:-] =Blockhead

I:(=Botox smiley

:-}X =Bow Tie-Wearing

%-6 =Brain Dead

:-(=) =Bucktoothed

:-E =Bucktoothed Vampire

:-#| =Bushy Mustache

})i({ =Butterfly

}:-X =Cat

q:-) =Catcher

C=:-) =Chef

;-(=Chin up

*<<<<+ =Christmas Tree

:-.) =Cindy Crawford

*<):o) =Clown

:-8(=Condescending Stare

%) =Confused

H-) =Cross-Eyed

:`-(=Crying

:*(=Crying softly

&:-) =Curly Hair

:-@! =Cursing

>:-> =Devilish

%-} =Dizzy

:3-] =Dog

:*) =Drinking every night

:#) =Drunk

<:-l =Dunce

:-6 =Eating Something Spicy

(:-| =Egghead

5:-) =Elvis

:-} =Embarrassed Smile

0|-) =Enjoying the Sun

>:) =Evil

>-) =Evil Grin

G(-'.'G) =Fighting Kid - it's a straight-on smiley

l:-O =Flat Top Loudmouth

:-! =Foot in Mouth

=:-H =Football player

:-W =Forked Tongue

%*@:-) =Freaking Out

/:-) =Frenchman with a beret

8) =Frog

: =Fuzzy

:} =Fuzzy With a Mustache

;S =Gentle warning, like "Hmm? What did you say?"

~~:-(=Getting Rained On

8*) =Glasses and a Half Mustache

}:-) =Hair Parted in the Middle Sticking up on Sides

:-}) =Handlebar Mustache

:-' =Has a Dimple

:(#) =Has Braces

:-`| =Have a Cold

/;-) =Heavy Eyebrows - Slanted

l^o =Hipcat

(_8(|) =Homer Simpson

*^_^* =Huge Dazzling Grin - it's a straight-on smiley

:0 =Hungry

o[-<]: =I am a skater

? =I don't understand what you mean

? =I have a question

?4U =I have a question for you

:-* =Kiss on the cheek

:p =Kitty with tongue hanging out

:(#) =Laughing like crazy

(-: =Left Hand

>;-> =Lewd Remark

:-9 =Licking Lips

;-, =Like, Duh

8:-) =Little Girl

%+{ =Lost a Fight

X-(=Mad

&-l =Makes Me Cry

:-(*) =Makes Me Sick

:-S =Makes No Sense

@@@@:-) =Marge Simpson

<33 =Meaning "heart or love" (more 3s is a bigger heart)

^^ =Meaning "read line" or "message above"

<3 =Meaning "sideways heart" (love, friendship)

<s> =Meaning "smile"

s =Meaning "smile"

w =Meaning "wink"

@|-) =Meditating Smiley

#:-) =Messy Hair

8(:-) =Mickey Mouse

:-{ =Mustache

:-3 =Mustache (Handlebar Type)

{:-{)} =Mustache and Beard

:-# =My Lips Are Sealed

(-) =Needs Haircut

):-(=Nordic

:/) =Not Amused

=8> Penguin

:^) =Personality

3:] =Pet Dog

:8) =Pig

:---) =Pinocchio

P-(=Pirate

3:[= Pit bull

}:^#) =Pointy Nosed

+<:-) =Pope

:-t =Pouting

+:-) =Priest

X:-) =Propeller Head

?-) =Proud of black eye

=:-) =Punk

=:-(=Punk Not Smiling

<(-'.'-)> =Puppy dog

:-r =Raspberry

(((((:-{= =Rave Dude

:-C =Real Unhappy

~:-(=Really Bummed Out

:-)) =Really Happy

([(=Robocop

[:] =Robot

@};--- =Rose

3:*> =Rudolph the red nose reindeer

:-(=Sad

:-d =Said with a Smile

:-y =Said with a smile

M:-) =Saluting

*<|:-) =Santa Claus

:-> =Sarcastic

:-@ =Screaming

$__$ =Sees Money

:-i =Semi-Smile

8-0 =Shocked

+-(=Shot Between the Eyes

:-V =Shouting

:O =Singing

~:-P =Single Hair

:-/ =Skeptical

':-/ =Skeptical again

:-7 =Skeptical variation

):-) =Smiley with Hair

:-, =Smirk

;^) =Smirking

~~~~8} =Snake

:-(<| =Standing Firm

=%-O =Stared at Computer Way Too Long

(8-{)} =Sunglasses, Mustache, Beard

/8^{~ =Sunglasses, Mustache, Goatee

:0 =Surprised

`:-) =Sweating

,:-) =Sweating on the Other Side

:-0 =Talkative

&-| =Tearful

-(:)(0)=8 =Teletubby

:-)--- =Thin as a Pin

:- ? =Tongue Sticking Out

:-& =Tongue Tied

:-a =Tongue Touching Nose

!#!^*&:- =Total Head Case

}(:-(=Toupee Blowing in Wind

<:>== =Turkey

x:-/ =Uncertain

=):-) =Uncle Sam

:-| =Unfazed
~~~~

|:-) =Unibrow

|:-| =Unyielding

:-[=Vampire

:-))) =Very Happy

%') =Very Tired

(:-(=Very Unhappy

:-< =Walrus

@:-) =Wavy Hair

{(:-) =Wearing a Toupee

[:-) =Wearing a Walkman

8-) =Wearing Contacts

B-) =Wearing Glasses

:-{} =Wearing Lipstick

]-I = Wearing Sunglasses

{:-) =Wears a Toupee

:-1 =Whatever

:-" =Whistling

;^? =Wigged Out

'-) =Winking

,-) =Winking Happy

;-) =Winking variation

8<:-) =Wizard

:-)8 : =Woman

:~) =Wondering

l-O =Yawning

|^o =Yawning or Snoring

:-(0) =Yelling

=8-0 =Yikes!

$-) =Yuppie

Text-tionary (Numbers)

They... not just for math... anymore.... They mean much more.... 14AA41(1 for all and all for 1)

Numbers

.02 = My (or your) two cents worth

121 =One-to-one (private chat initiation)

143 =I love you

14AA41 =One for all, and all for one

10X =Thanks

1CE =Once

1DR =I wonder

2EZ =Too easy

2G2BT =Too good to be true

2M2H =Too much too handle

2MI =Too much information

2MOR =Tomorrow

2NTE =Tonight

4 =Short for "for" in SMS

404 =I don't know

411 =Meaning "information"

459 =Means I love you (ILY is 459 using keypad numbers)

4COL =For crying out loud

4EAE =Forever and ever

4NR =Foreigner

^5 =High-five

6Y =Sexy

7K =Sick

831 =I love you (8 letters, 3 words, 1 meaning)

86 =Over

9 =Parent is watching

www.Text-tionary.com

For even more info... go to www.text-tionary.com find out what's new in the texting world, where you can find your very own Text-tionary, learn a new text for the month!!!!! Talk to me, do you have a texting gone wrong story... we want to hear about texting stories, the good, the bad, as well as the ugly. Submit your texting stories...... www.text-tionary.com